Biological & Behavioural H

Ibrahim El- Sayed Shehata

Biological & Behavioural Habits of the Tomato Leaf Miner Tuta absoluta

(Geliicidae:Lepidoptera)

LAP LAMBERT Academic Publishing

Imprint

Cover image: www.ingimage.com

Publisher:
LAP LAMBERT Academic Publishing
is a trademark of
International Book Market Service Ltd., member of OmniScriptum Publishing Group
17 Meldrum Street, Beau Bassin 71504, Mauritius

Printed at: see last page
ISBN: 978-613-7-32747-0

NAME: IBRAHIM E. SHEHATA

[National Research Centre, Egypt]

Population density and infestation rate of the tomato leaf miner *Tuta absoluta* (Geliicidae:Lepidoptera)

Preface

History of *Tuta absoluta*

Tuta absoluta (Meyrick, 1917) was first described in Peru in 1917 and is considered one of the most devastating pests for the tomato crop in South America. It is defined as tomato leaf miner. This moth belongs to of the family Gelechiidae (Lepidoptera) and is one of the most serious pests of tomato (Solanum lycopersicum L.) (Solanaceae) **(Desneux et al. 2010).** It also attacks other cultivated Solanaceae, such as potato, eggplant, peppers and tobacco **(Tumuhaise et al. 2016),** although **Potting et al., (2013)** disputed peppers as a host. Tomato leaf miner (TLM) also feeds on various solanaceous weeds **(Chidege et al. 2016).**

Tuta absoluta was originally described by **Meyrick, (1917)** as *Phthorimaea absoluta*, based on individuals collected from Huancayo (Peru).

Povolny, (1964) described it as *Scrobipalpula* in 1964, and transferred *absoluta* to this genus. **Clarke, (1965)** transferred *absoluta* to *Gnorimoschema* as a new combination.

Becker, (1984) included *absoluta* in *Scrobipalpula* in Heppner's Atlas of Neotropical Lepidoptera, but did not list it as a new combination.

Povolny, (1987) described *Scrobipalpuloides* as a new genus and transferred *absoluta* to this genus as a new combination. The same author in **1994** transferred *absoluta* from *Scrobipalpuloides* to *Tuta* as a new combination and identified this pest as *Tuta absoluta*.

Geographical distribution.

The tomato leaf miner *T. absoluta* (Meyrick) originated from South America and is known as a tomato pest in countries of this continent.

T. absoluta was recorded in different countries as Venezuela **(Fernandez and Montagne, 1990a)**, Argentina **(Giganti and Graciela-Gonzalez, 1993)**, Bolivia **(Zhang, 1994)**, Ecuador **(Povolny, 1994)**, Brazil, Uruguay and Chile **(Pastrana, 2004)**, Colombia **(Colomo and Berta, 2006)**, Panama **(Russell IPM, 2009; USDA,**

2011), Paraguay (**EPPO, 2011a**). According to **Campos et al., (2017)** Country specific first reports of *Tuta absoluta* outside of South America listed as follow.

Country	year	Reference
Spain (including Balearic islands)	2006	(Urbaneja et al. 2007)
Albania	2008	(EPPO 2009a)
Algeria	2008	(Guenaoui 2008)
France	2008	(EPPO 2009c)
Italy	2008	(Viggiani et al. 2009)
Morocco	2008	(EPPO 2008)
Tunisia	2008	(EPPO 2009g)
Belgium	2009	(VanDamme et al. Undated)
Bulgaria	2009	(EPPO 2010a)
Croatia	2009	(EPPO 2011b)
Cyprus	2009	(EPPO 2010b)
Germany	2009	(EPPO 2010c)
Greece (including Crete)	2009	(Roditakis et al. 2010)
Israel	2009	(EPPO 2010d; Seplyarsky et al. 2010)
Libya	2009	(EPPO 2011d)
Lithuania	2009	(Ostrauskas and Ivinskis 2010)
Malta	2009	(EPPO 2009d)
Portugal (including Azores)	2009	(EPPO 2009e)
Slovenia	2009	(Knapič and Marolt 2009)
Switzerland	2009	(EPPO 2009f)
The Netherlands	2009	(NPPO 2009)
United Kingdom (Including Guernsey)	2009	(EPPO 2009b)
Austria	2010	(EPPO 2011a)
Bosnia and Herzegovina	2010	(Đurić et al. 2012)
Egypt	2010	(EPPO 2011d)
Hungary	2010	(EPPO 2010e)
Iraq	2010	(Abdul Razzac et al. 2010)
Kosovo	2010	(EPPO 2010f)
Kuwait	2010	(EPPO 2011d)
Montenegro	2010	(Hrncic and Radonjic 2011)

Romania	2010	(Keresi et al. 2010)
Southern Russia	2010	(Izhevskya et al. 2011)
Saudi Arabia	2010	(Almatni 2010; EPPO 2011d)
Serbia	2010	(Toševski et al. 2011)
Sudan	2010	(EPPO 2011d; Mohamed et al. 2012)
Syria	2010	(Almatni 2010)
Turkey	2010	(Kılıç 2010)
Bahrain	2011	(EPPO 2011d)
Cape Verde (Santiago Island)	2011	(Duarte 2013)
Cayman Island	2011	(USDA–APHIS 2011)
Ethiopia	2011	(Pfeiffer et al. 2013; Retta and Berhe 2015)
Georgia	2011	(EPPO 2011c)
Iran	2011	(Banamieri and Cheraghian 2011)
Jordan	2011	(EPPO 2011d)
Lebanon	2011	(EPPO 2011d)
Panama	2011	(EPPO 2012a)
Qatar	2011	(EPPO 2012b)
Canary islands	2012	(Polaszek et al. 2012)
Niger	2012	(Pfeiffer et al. 2013)
Senegal	2012	(Brévault et al. 2014)
United Arab Emirate	2012	(EPPO 2013b)
Armenia	2013	(Schaub 2013)
Czech Republic	2013	(EPPO 2013a)
Yemen	2013	(EPPO 2013c)
Benin	2014	Unconfirmed
Costa Rica	2014	(EPPO 2014a)
Ivory Coast	2014	Unconfirmed
Ghana	2014	Unconfirmed
Giunea-Bissau	2014	Unconfirmed
Guinea	2014	Unconfirmed
India	2014	(Kalleshwaraswamy et al. 2015; Sridhar et al. 2014)
Kenya	2014	(IPPC 2014)
Liberia	2014	Unconfirmed
Mali	2014	Unconfirmed

Pakistan	2014	unconfirmed
Oman	2014	Unconfirmed
Sierra Leone	2014	Unconfirmed
South Sudan	2014	Unconfirmed
The Gambia	2014	Unconfirmed
Togo	2014	Unconfirmed
Ukraine	2014	(EPPO 2014b)
Afghanistan	2015	(Russell-IPM 2016)
Eritrea	2015	(Tonnang et al. 2015)
Mayotte Island	2015	(EPPO 2016)
Nigeria	2015	(FAO 2015)
Rwanda	2015	(FAO 2015)
Tanzania	2015	(Materu et al. 2016)
Turkmenistan	2015	(Bratu et al. 2015)
Uganda	2015	(Tumuhaise et al. 2016)
Bangladesh	2016	(Hossain et al. 2016)
Botswana	2016	(Tebele 2017)
Burkina Faso	2016	(Son et al. 2017)
Namibia	2016	(Kaira 2017)
Nepal	2016	(NARC 2016)
South Africa	2016	(IPPC 2016)
Zambia	2016	(IPPC 2016)
Burundi	2016	unconfirmed
Azerbaijan	2016	unconfirmed
Democratic Republic of the Congo	2016	unconfirmed
French Guiana	2016	unconfirmed
Guyana	2016	unconfirmed
Kyrgyzstan	2016	unconfirmed
Malawi	2016	unconfirmed
Suriname	2016	unconfirmed
Tajikistan	2016	unconfirmed
Uzbekistan	2016	unconfirmed
Zimbabwe	2016	unconfirmed
Mozambique	2017	(IPPC 2017)

Fig. 1:

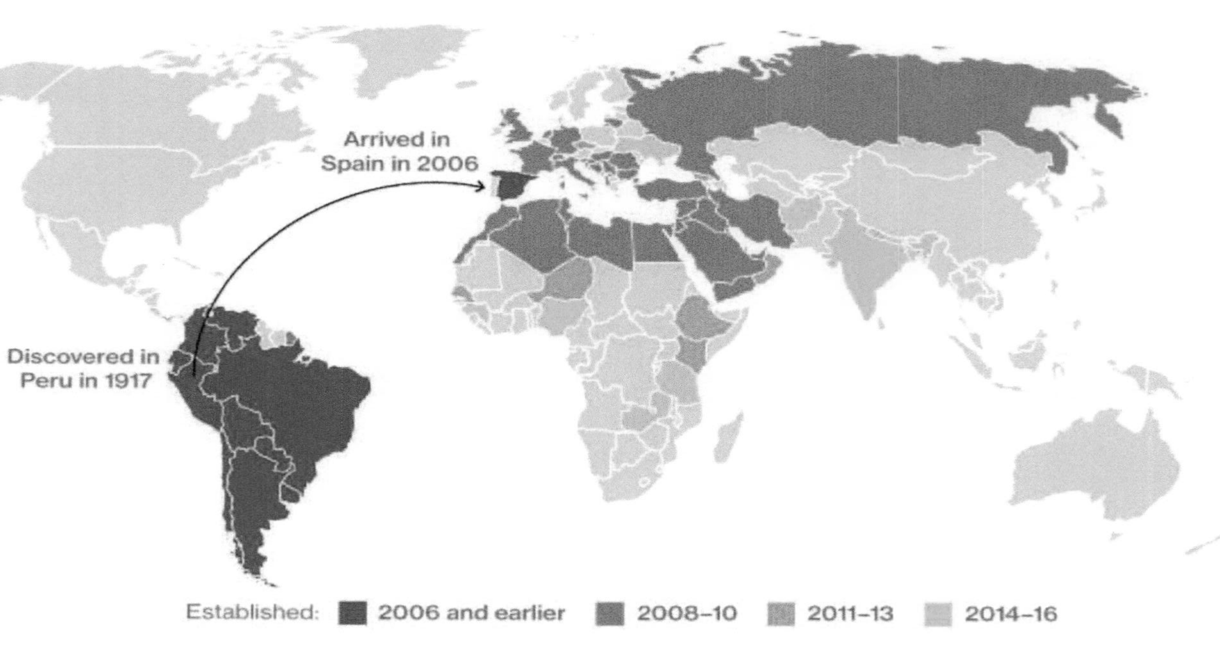
The Pest That's Infesting Tomato Crops
Since 2008, Tuta absoluta has spread to 15 African countries
and driven up costs for both farmers and consumers
Arrived in
Spain in 2006
Discovered in
Peru in 1917
Established: 2006 and earlier 2008–10 2011–13 2014–16
Sources: European and Mediterranean Plant Protection Organization (2005), Office of International Research,
Education, and Development, Virginia Tech

Biology and Life cycle of *T. absoluta*

T. absoluta is a micro lepidopteran moth with high reproductive potential, adults can be detected around the year. *T. absoluta* is a night active moth with a high reproduction capacity that allows the pest population to increase rapidly. Females can lay up to 260 eggs **(EPPO, 2005)** on the underside of leaves or on stems and to a lesser extent on fruits. After hatching in the morning larvae normally start mining within one hour. The life cycle of *T. absoluta* passes through four developmental stages: egg, larva, pupa and adult as shown in **fig. (2)**.The environmental conditions such as temperature and relative humidity are factors influencing insect physiology and behavior. Temperature has a direct influence on the insect activity and rate of development. The rate of development and life cycle is based on the accumulation of heat measured in physiological rather than chronological time, **(Varges, 1970).**

Salama *et al.*, (2014) described the biological stages as follow

Egg stage

The newly deposited eggs are oval in shape and creamy white in color then turn to yellow and finally black before hatching. The incubation period was 14.65 ± 0.9 days at 15 °C, then decreased to 3.7 ± 0.16 days as the temperature increased to 30 °C.

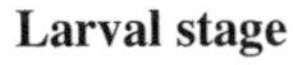

Larval stage

The larvae of this moth characterized by the presence of two dark spots on the porthorax.

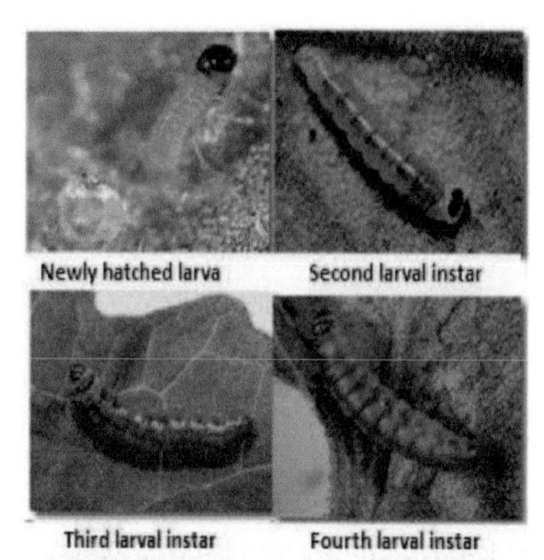

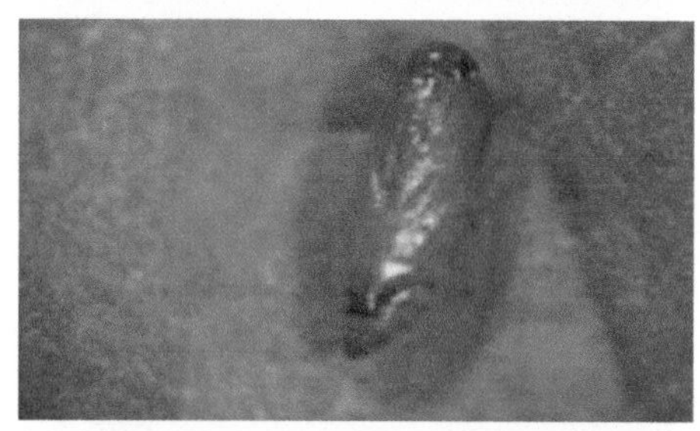

Fig. 2. Different stages of *T. absolut*

T. absoluta passes through four larval instars. Just after hatching the larvae are creamy yellow, then they feed and become greenish in color and the dorsal region turns to reddish when they are close to pupate. The larval duration was 31.98±1.7 days at 15 ˚C, increase of temperature shortened significantly to being 9.2 ± 0.21 days at 30 ˚C.

Pupal stage

After development of 4th larval instar the insect pupates, and the pupa has a cylindrical shape, greenish in color then turns to brown. The mean duration of the pupal stage was 22.13 ± 0.53 days at 15 ˚C and 6.48 ± 0.19 days at 30 ˚C.

Adult stage

Following pupation, the moth emerges to give male and female. Both sexes can be distinguished through their size, where the females are wider and more voluminous than the males. The abdomen of the female is brown in color but it is creamy in the male. The male longevity was 21.6 ± 0.8 and 6.7 ± 0.2 days at 15 and 30 ˚C, respectively, while the female longevity lasted longer being 25.8 ± 1.9 at 15 ˚C and it decreased to 14.3±0.2 days at 30 ˚C

Behavioral habits

T. absoluta is nocturnal in habits. Adults are most active at dusk and dawn and rest among leaves of the host plant during the day, showing greater morning-crepuscular activity. The duration of the life cycle depends on the environmental conditions, in particular the temperature, ranging from 76.3 days at 14°C to 23.8 days at 27°C **(Barrientos *et al.*, 1998).** Also there may be an overlap in generations and all the stages of the cycle are found in field conditions **(Souza and Reis 1992).**Temperature variations directly influence the development rate of insects and indirectly influence their development through effects on the host plant, and so, the plant quality as food resource **(Karban and Baldwin 2007).** Many investigations were carried out to determine the insect behavior towards different factors with effects of the environmental conditions on the biological aspects. **Fernandez and Montagne, (1990a)** reported that the egg stage lasted 4.4 to 5.8 days at 24.6°C and larval stage development was completed in 11 to 15 days at 24.09 °C and then adult males emerged in 7 to 8 days and females in 6 to 8 days at 26.3°C. The average pre-oviposition period for females was 2.4 ± 0.61 days and females can lay eggs for more than 20 days; however, 72.3 % of eggs were deposited during the first 5 days and 90 % in the first 10 days. While, **Barrientos *et al.*, (1998)** showed that the duration of the developmental cycle greatly depends on the environmental conditions, with average development time of 76.3, 39.8 and 23.8 days at 14, 20 and 27 C°, respectively.

Mating behavior

The sex in *T. absoluta* is separated and the male is attracted to female by the response to pheromones **(Uchoa-Fernandes *et al.*, 2005). Santos-Silva, (2008).** Reported that males mate more times (up to 12 times in their lifetime) than females. In laboratory studies conducted in Venezuela, adults mated up to 16 times during their lifetime **(Fernandez & Montagne, 1990).** However, data from laboratory mating studies may not reflect the true number of matings taking place in the field because **Vargas, (1970)**, found no information on the number of matings for *T.*

absoluta under field conditions and showed both males and females exhibit a strong phototactic response. Egg deposition takes place throughout the day, but peak oviposition occurs at night. Observations of **Salama et al., (2015)** on mating behavior showed that the best time for mating was in the early morning at 7.30 a.m. and the mating process lasted for 5.14± 1.15 hours. These observations agree with those of **Imenes *et al.*, (1990)** who studied the biology and behavior of *T. absoluta* in the laboratory and observed that the mating activities started at 7 a.m approximately, and duration of mating took about 4.49 hours. In this concern, **Hickel *et al.*, (1991)** studied the mating behavior of *T. absoluta* in the laboratory and reported that the male mating behavior can be divided into two phases: long-range female location and short-range courtship. Long-range female location includes behavior that eventually leads to the arrival of males in the vicinity of females. Short-range courtship behavior focus on interactions between the genders that eventually lead to mating. Duration of copula is variable, sometimes taking 2 to 3 hours or extending to as much as 6 hours. Both genders mate multiple times. Also, **Ucho^a-Fernandes *et al.,* (1995)** showed that, females of *T. absoluta* mate only once a day and are able to mate up to six times during their lifespan, with a single mating lasting about 4–5 hours. **Miranda-Ibarra, (1999)** reported that the greatest number of males was captured in pheromone traps during the period 7 to 11 a.m. suggesting that this is the best time when males are searching for calling females. Similarly, **Marina *et al.,* (2014)** stated that mating always began during the first hour of the photophase, and mating pairs took from a few minutes to 6 h 40 min to uncouple and 76 % of females laid eggs on the same days they mated. The variation in copulation time may be proportional to completely transfer the spermatophore from male to the female (**Ouye *et al.*, 1965; Seth *et al.*, 2002**) or may be proportional to spermatophore size **(Franco *et al.*, 2011).**

Female & Male

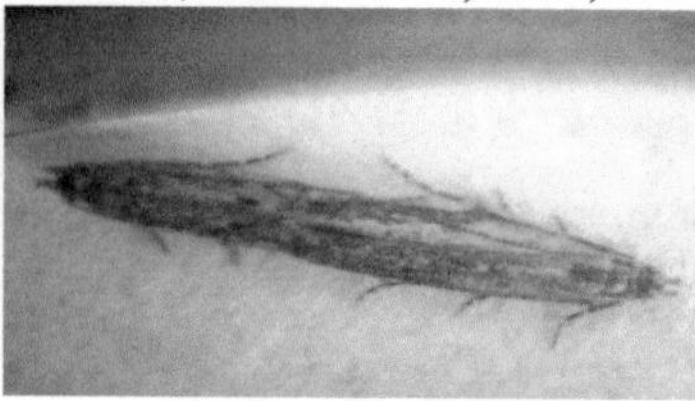

Mating position

Egg production

The insect mainly presents nocturnal habits and adults usually remain hidden during the day showing greater morning-crepuscular activity with adults dispersing among crops by flying among a range of species within the Solanaceae. The female of *T. absoluta* lays up to 260 eggs on both sides of leaves or stems of its host during its life and females mated only once a day and are able to mate up to six times during their lifespan, with a single mating lasting 4-5 hours **(Ucho^a-Fernandes *et al.*, 1995)** *T. absoluta* female laid about 60.56% of her egg on the upper side surface of tomato leaf, while the lowest (0.85%) was laid on tomato stem. The highest number of eggs of 233.75±14.42 was laid at 20.5±2°C and 55±5% R.H., whereas the lowest number of eggs (177.5±9.26) was laid at 32.0±2°C and 40±5% R.H **(Shiberu and Getu, 2017). Salama *et al.*, (2014)** reported that the highest egg production was obtained at 20 ˚C being 162 ± 30.94 eggs/ female, this decreased to 159 ± 35.5 eggs / female at 15 ˚C. At higher temperatures, the mean production was 128.9 ±31.7 and 115.3 ± 23.64 eggs / female after rearing in 25 and 30 ˚C, respectively. The highest egg deposition was tenth, seventh and fifth days after the female emergence at the 15, 20, 25 and 30 ˚C, respectively. The average pre-oviposition period for females was 2.4 ± 0.61 days and females can lay eggs for more than 20 days; however, 72.3 % of eggs were deposited during the first 5 days and 90 % in the first 10 days. the egg stage lasted 4.4 to 5.8 days at 24.6°C **Fernandez and Montagne, (1990a)** while, **Marina *et al.*, (2014)** reported that the most of the females laid eggs on the same days they mated.

The females deposited their eggs on all parts of host plants; leaves, leaf neck, stem and sepals. The leaves were more attractive to females and the lower leaf surface appeared to be more infested than the upper surface, after egg hatching, the larvae of *T. absoluta* penetrated tomato leaves, feed on leaf parenchyma tissues forming irregular leaf mines that get longer and wider as the larvae continue to feed **(Salama et al., 2015)** causing reduction in photosynthesis causing stunted in plant growth. **Leite *et al.*, (2004)** demonstrated that there was a preferential deposition of *T. absoluta* eggs on the apical leaves of the tomato plant.

Riquelme, (2009) showed that *T. absoluta* prefers to lay eggs on the leaves (both sides) and on aerial parts of the plant and observed no significant differences in the vertical distribution of eggs on the plant. However females tended to concentrate their egg laying activity on the upper third of the plants after the third week of planting. The apical part of the plant was more attractive to the females compared to the median and basal parts **(Liliana *et al.*, 2010; Proffit *et al.*, 2011; Cherif *et al.*, 2013)**. The newly hatched larvae tend to orient itself and settle in the dark zones. They behave like a borer when it feeds on the leaf surface and then penetrate within the leaf tissues within 30-37 minutes **(Salama et al., 2015).** This is similar with the results of **Cuthbertson *et al.*, (2013)** who reared *T. absoluta* and showed that the first instar larvae fed on the leaf surface for approximately 82 minutes before becoming fully submerged inside the leaf. The larvae of *T. absoluta* mainly depend on green tissues of leaves in feeding and prefer to feed on stems, buds and the calyx over tomato fruit which can affect the plant photosynthetic capabilities **(FERA, 2009).**

The senses of taste and smell in lepidopterous larvae are important in host plant selection and the acceptance of the larva of *T. absoluta* to a host plant is due probably to chemical volatiles rather than purely physical factors as reported by **Torres *et al.*, (2001)** and **Proffit *et al.*, (2011)**. In this concern, **Kamel, (1969)** found that the larvae of *Spodoptera littoralis* bear gustatory receptors that are localized on the labrum epipharynx.

Life cycle

The duration of the developmental cycle greatly depends on the environmental conditions. **Chiang (1985)** denominated "optimal range" to the temperature going from the lower to the upper threshold, where the development is directly proportional to temperature. Outside these limits, activity decreases to almost a standstill without necessarily causing death.

Adults are 6–7 mm in length and present filiform antennae and silver to grey scales **(Coelho & Franca, 1987).** Black spots are present on anterior wings, and the females are wider and more voluminous than the males. The pest mainly presents nocturnal habits, and adults usually remain hidden during the day, showing greater morning-crepuscular activity with adults dispersing among crops by flying.

After mating eggs are laid on the aerial parts of plant and hatch to first larval instar young larvae penetrate leaves, aerial fruits or stems, on which they feed and develop **(EPPO, 2005).** *T. absoluta* passes through four larval instars **(Verges, 1970).** The duration of larval stages depend on the temperature where the development stops at 6-9 °C. The average development time was 76.3 days at 14 °C, 39.8 days at 20 C°, and 23.8 days at 27 °C, so the last degree appeared to be the optimum for *T. absoluta* development **(Barrientos *et al.*, 1998)**. Fully-fed larvae (The fourth larval instar) usually drop to the ground on a silk thread and pupate in the soil, although pupation may also occur on leaves and in mines **(Van-Deventer, 2009).** Pupae are cylindrical in shape and greenish when just formed becoming darker in color as they are near adult emergence.

In the optimum conditions *T. absoluta* has 10 to 12 generations per year, since the hatching of its eggs takes 4–5 days, the four larval instars takes 13–15 days, pupae phase takes 9–11 days, and yet the one generation total duration is about 26-31days **(Desneux *et al.*, 2010)** Adults are active at night and females lay eggs on the aerial parts of the host 85 Berxolli and Shahini 2017 plants. Adults are 6–7 mm in length and present filiform antennae and silver to grey scales **(Coelho *et al.*, 1987).** The maximal lifetime fecundity is 260 eggs per female **(Ucho^a-Fernandes et al., 1995).** Larvae do not enter in diapauses when food is available **(EPPO, 2005).** The potential number of generations calculated in the Algiers region in 2009 is 8.3. In 2010 it is equal to 7.72 **(Mahdi & Doumandji, 2011).** In the same direction, **Abolmaaty, (2010)** evaluated how temperature influence the annual generation in four governorates in Egypt (EL-Beheira, Giza, Qena and Fayoum governorates) his obtained results indicated that the population of *T. absoluta* in Qena governorate gave the highest number of generation as compared to other locations (EL Beheira, Giza and Fayoum governorates) under current climate, the number of generations of *T. absoluta* exhibited 11, 12, 12 and 13 at El-Beheira, Giza, Fayoum, and Qena, respectively. While, 13.6 generations were recorded per year in Assiut governorate,

Egypt **(Samy, 2011). Shehata, (2015)** expected the annual generations of *T. absoluta* to be 3.86, 5.37, 8.04 and 11.67 generations per year at rearing in 15, 20, 25 and 30 ˚C under laboratory conditions. **Salama *et al.*, (2014)** designed the life table of *T. absoluta* as follow.

Life table of *Tuta absoluta* under controlled temperature

Biological aspects	Durations / days ± SD at temperatures			
	15 ˚C	**20 ˚C**	**25 ˚C**	**30 ˚C**
Eggs	14.65 ± 0.9	8.7 ± 0.4	5.61 ± 0.2	3.7 ± 0.16
Larvae	31.98 ± 1.7	21.4 ± 1.1	14.56 ± 0.2	9.2 ± 0.21
Pupae	22.13 ± 0.53	16.58 ± 1.2	10.93 ± 0.31	6.48 ± 0.19
Longevity of Male	21.6 ± 0.8	16.3 ± 0.4	10.1 ± 0.7	6.7 ± 0.2
Longevity of Female	25.8 ± 1.9	21.3 ± 0.9	14.3 ± 0.2	11.9 ± 0.18
Life cycle of male	90.36 ± 1.6	62.98 ± 7.8	41.2 ± 3.7	26.08 ± 0.34
Life cycle of female	94.56 ± 1.2	67.98 ± 1.6	45.4 ± 0.61	31.28 ± 0.44
Pre-oviposition	4.6 ± 0.5	4 ± 0.63	2.9 ± 0.83	2.3 ± 0.65
No. of eggs/ Female	159 ± 35.5	162 ± 30.94	128.9 ± 31.7	115.3 ± 23.64

Fig. 3. Life cycle of *T. absoluta*

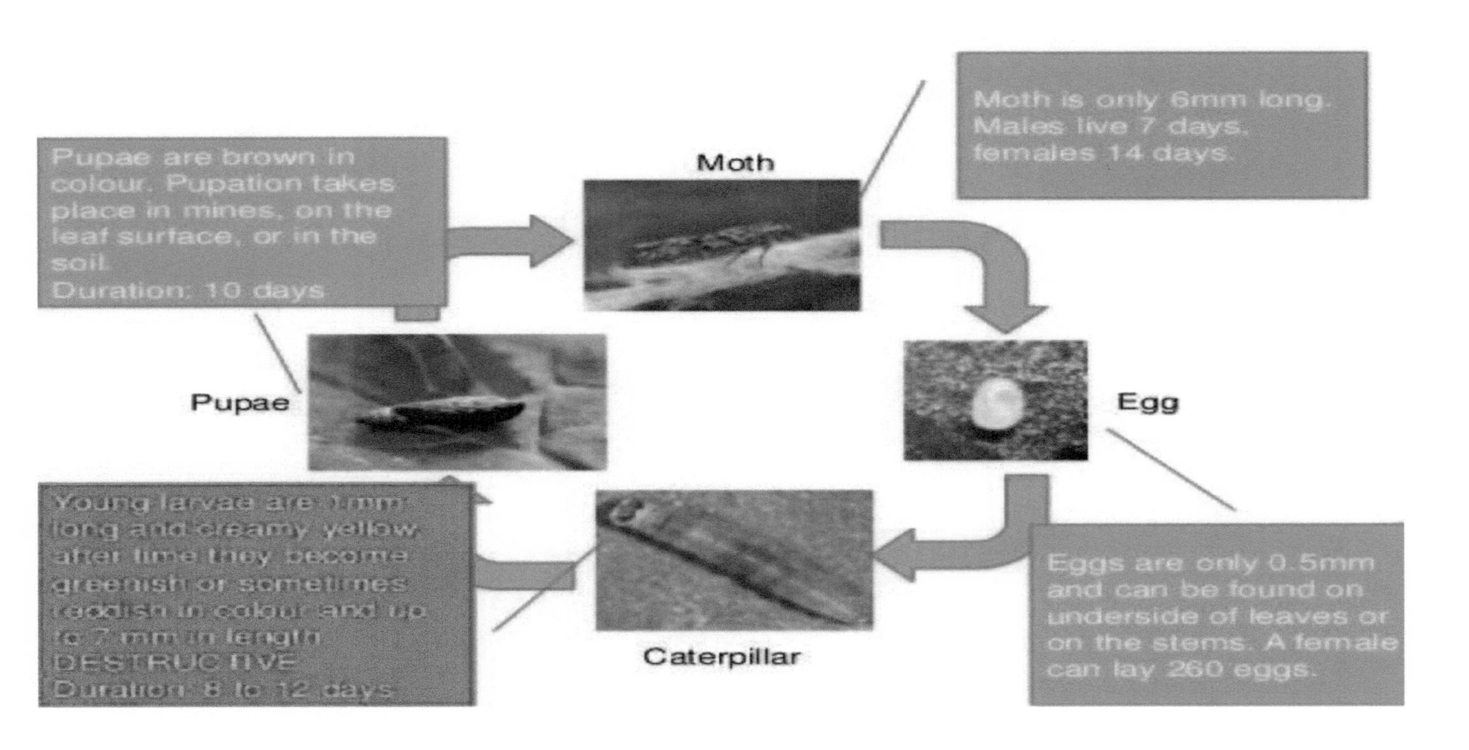

Damage of *T. absoluta*

The tomato plants are infested at any developmental stage with females ovipositing preferentially on leaves, tender portion of the stem, developing and mature fruits causing fruit malformation and fruit rot. The pest mainly presents nocturnal habits and adults usually remain hidden during the day showing greater morning crepuscular activity with adults dispersing and flying among crops. This pest infests Solanaceous crops mainly tomato, potato, pepper and eggplant.

After egg deposition, larvae penetrate the leaves, stems or fruits causing mines and galleries. No tomato cultivars are entirely resistant to this insect, but not all cultivars are equally susceptible. Similar observations were recorded by **Borgorni *et al.*, (2003); Oliveira *et al.*, (2009); Proffit *et al.*, (2011); De Oliveira *et al.*, (2012)** and **Cherif *et al.*, (2013).** The tomato leaf miner prefer to lay eggs on the leaves (both sides), larvae normally hatch from the eggs in the morning and penetrate the leaf surface causing irregular mines resulted from feeding on green leaf tissues. The movement of larvae of *T. absoluta* observed and the study reported that, the larvae have been observed walking on leaves outside of their mines. This behavior might be related to the temperature inside the mine, the depletion of food, and/or the accumulation of fecal material. Leaf mines have an irregular shape and may later become necrotic and the galleries in the stems may alter the general development of the plant, the serpentine-shaped mines increase in length and width as the larva develops and feeds **(Torres *et al.*, 2001).** Also, the fruits can be attacked as soon as they are formed, and the galleries bored inside them can be invaded by secondary pathogens leading to fruit rot as reported in **(Vargas, 1970; EPPO, 2005).**

The tomato leaf miner prefer to lay eggs on the leaves and they will oviposit on other aerial parts of the plant such as shoots, stems, flowers and green fruit underneath the sepals that form the calyx and observed no significant differences in the vertical distribution of *T. absoluta* eggs on tomato plants, however females tended to concentrate their egg laying activity on the upper third of the tomato plants after the third week of planting **(Riquelme, 2009)**. Tomato plants can be attacked at any developmental stage, with females ovipositing preferentially on leaves (73%), and to lesser extent on leaf veins and stem margins (21%), sepals (5%) or green fruits (1%) **(Estay, 2000).** Oviposition was found possible on unripe tomatoes only **(Monserrat, 2009).** The field investigations of **Salama et al., (2015)** recorded that the eggs were recorded on all parts of the plant except flowers and fruits where no eggs were detected. The leaves were more attractive to females followed by sepals, leaf neck and stem where the percentages of deposited eggs were 96.45, 2.3, 0.7 and 0.61 %, respectively. The lower leaf surface was more infested than the upper surface and the percentage of deposited eggs was 52.06 and 47.94 %, respectively. The apical parts showed to be more attractive to the female compared to the median and basal parts, where the average number of deposited eggs on the apical parts was 49.8 eggs (43.12%) distributed on leaves (96.6 %), leaves neck (1.2 %), stem (1 %) and sepals (1.2 %). Meanwhile, the average number of deposited eggs on the median parts was 38.9 eggs (33.67 %) distributed on leaves (96.14 %), leaves neck (0.52 %), stem (0.52 %) and sepals (2.82 %). On the other hand, the basal part of the tomato plant was less attractive where the average number of deposited eggs on it was 26.8 eggs (23.2 %)

distributed on leaves (96.64 %) and sepals (3.36 %) with no record of eggs on both stem and leaf neck.

In laboratory studies conducted in Venezuela, 96.8 % of a cohort of 94 eggs eclosed between 6 and 9 Am as reported by **Fernandez & Montagne, (1990)**. After hatching, young larvae penetrate the leaves, stems or tomato fruits on which they feed and develop, creating conspicuous mines and galleries. In tomato leaves, damages are caused through mine-formation within the mesophyll by feeding larvae, thus affecting the plant's photosynthetic capacity and consequently lowering tomato yield. Galleries in stems alter the general development of the plant and could cause necrosis. Fruits can be attacked as soon as they are formed, and the galleries bored inside them can be invaded by secondary pathogens leading to fruit rot. Finally, an important additional problem is that the pest directly feeds on the growing tip, thereby halting plant development. The pest affects tomatoes destined to fresh market as well as to processing, with larvae causing losses in its area of origin of up to 80–100% **(Vargas, 1970; Lo´pez, 1991 and Apablaza, 1992)**. Furthermore, feeding activity on fruits directly affects the visual aspect of harvested products.

Although larvae spend most of their life inside mines, second instars can leave the mines, thus exposing them to predation, well-timed application of pesticides and possibly parasitism, larvae have been observed leaving their mine and starting a new mine on a different part of the plant, when outside of the mines larvae move quickly and can also pull together new shoots using silk silken threads produced by specialized salivary glands. **(Fernandez & Montagne, 1990; USDA, 2011)**

The damage caused by this pest is severe; especially in young plants. The larva usually enters the fruit under the calyx and tunnels the flesh, leaving galleries clogged with frass that cause the fruit to drop or to rot on the vine. Larvae can also enter the fruit through the terminal end or through other fruit parts that are in contact with leaves, other fruits, or stems. Without adequate controls, infestations of *T.*

absoluta can result in 90 to 100 % loss of field-produced tomatoes **(Vargas, 1970; Estay, 2000).**

Affected fruit lose their commercial value. Larval damage to terminal buds in greenhouse-grown tomatoes in Argentina negatively affects plant architecture and can result in a significant reduction of fruit yield **(Botto, 2011b)**. Mature larvae purge themselves of food and shorten their body length. Larvae spin a silken cocoon where they transform into pupae. Pupae can be found attached to all plant parts (leaves, main stem, flowers and fruit) as well as in the soil **(Torres *et al.*, 2001).**

Gharekhani & Ebrahimi, **(2013)**. Focused on evaluating the damage of *T. absoluta* on eleven 45-day-old tomato cultivars under greenhouse condition. Larval mines on the leaves as well as the terminal bud damage were considered. Damaged leaves, active mines and damaged terminal buds were significantly different among the cultivars. Cluster analysis using SPSS software resulted in grouping the cultivars into four categories as relatively resistant, partially resistant, partially susceptible and susceptible. The host plant's growing characteristics (height and leaflet number) were assessed and likewise the weight of the resulted pupae. Differences in vulnerability of the cultivars showed that tomato cultivars possess resistant traits and the identification and utilization of these traits can give rise to resistant varieties.

T. absoluta is a very harmful leaf mining moth with a strong preference for tomatoes. It also occurs on eggplants, sweet peppers as well as potatoes and various other cultivated plants. It also occurs on weeds of Solanaceae family (Solanum nigrum, Datura spp.). It can cause 50-100% yield reduction on tomato crops and its presence may also limit the export of the product to several destinations. Prevention and proper management of the pest are crucial. Chemical control often fails due to the resistance of this pest against many pesticides, but also because a big part of its development takes place inside the plant or the soil, out of reach of pesticides.

The symptoms of infestation with *T. absoluta* on fruits and leaves of tomatoes photographed as in **Fig. (4)**. Most distinctive symptoms are the blotch-shaped mines in leaves, sepals and fruits. Inside these mines both the caterpillars and their dark frass can be found. In case of serious infection, leaves die off completely.

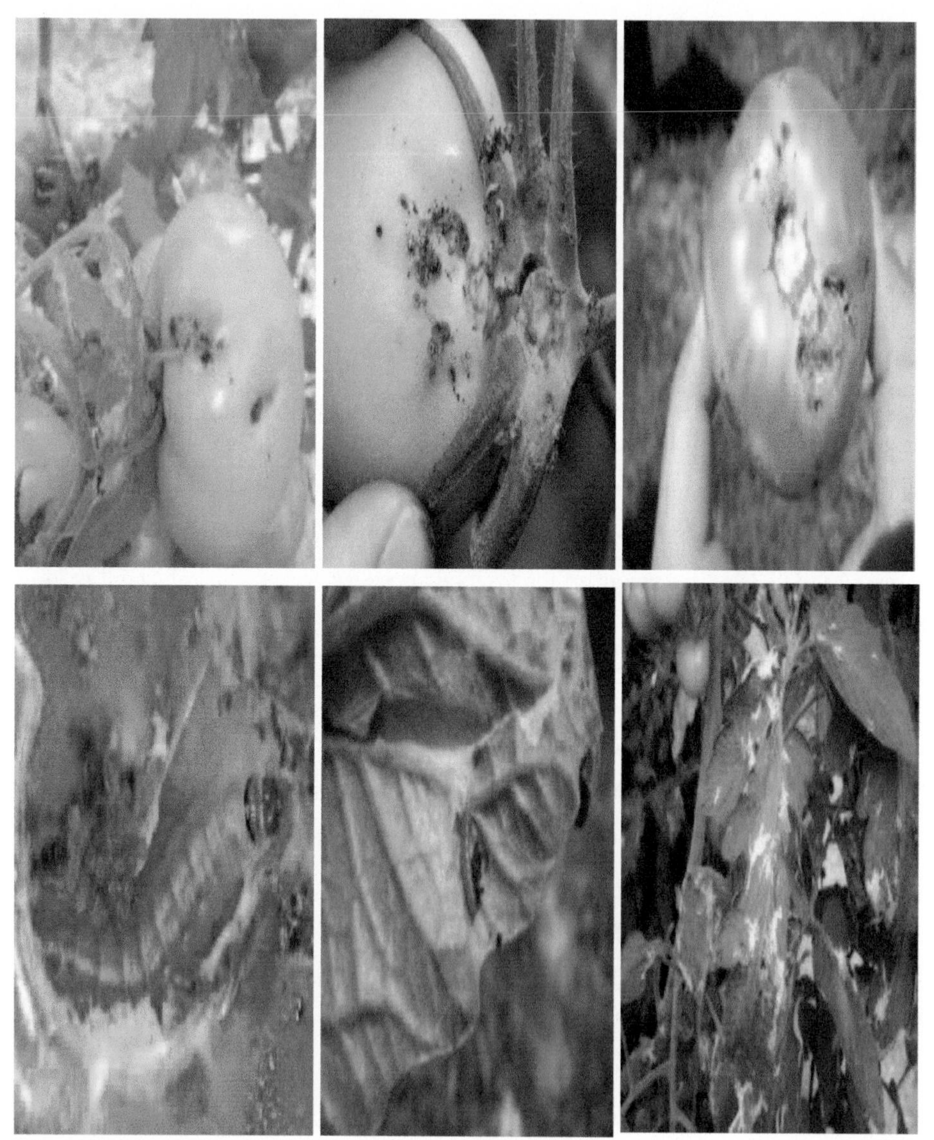

Fig. (4). Symptoms of infestation with *T. absoluta*

Tomato crops exposed to attack by different insect species that reduce the quality and quantity of crops production such as leaf miner insects (*Tuta absoluta* & *Liriomyza tifolii*) the larvae of these pests attack different vegetable crops as tomato, potato, eggplant, pepper and bean in all developmental stages, where the larvae prefer leaves, stems, flowers and especially fruits besides mining their leaves which can affect the plant photosynthetic capabilities causing high reduction in productivity 80-100% **(FERA, 2009).**

There are similarities in the mode of damage resulting from the infestations with larvae of both *T. absoluta* (Order: Lepidoptera) and other leaf miners (Liriomyza species) belonging to order Dipetra.

Leaf miners belonging to the genus Liriomyza are considered pests in many crops due to their damage to the leaves **(Robert 1999).** There are four leaf miner species which are common pests: the tomato leaf miner (*Liriomyza bryoniae*), the American serpentine leaf miner (*Liriomyza trifolii*), the pea leaf miner (*Liriomyza huidobrensis*) and *Liriomyza strigata*. All these under natural conditions, the larvae of these species are well parasitized by several natural enemies. Problems with leaf miner have increased as a result of the use of broad-spectrum pesticides: natural enemies are killed and the leaf miners develop resistance to these insecticides.

The leaf miner goes through six stages, namely egg, three larval stages, pupa and adult. The adult leaf miners are small, yellow and black colored flies. The larvae form mines in the leaves of plants. Pupation takes place mostly in the soil.

Larvae cause mines. This can lead to cosmetic damage, leaves drying out or even early defoliation. The latter may affect the yield. Female adults cause feeding marks where they feed. This gives cosmetic damage to the plants. Indirect damage occurs when fungi or bacteria enter the feeding areas.

The variation between damage caused by *T. absoluta* and *Liriomyza* species is diagnosed by the shape of leaf mines. The mines were regular at the presence of *T. absoluta* but irregular shape at *Liriomyza* sp. as shown in **fig. 5**

DAMAGE TO STEMS, LEAVES AND FRUITS

Fig. 5. Damage caused by *T. absoluta* and *Liriomyza*

Host preference

T. absoluta is polyphagous. It is one of the most important lepidopterous insect pests on tomato in both greenhouses and in the open field. This pest feeds on several solanaceous species, and preferentially on tomato *Solanum lycopersicum* **(Siqueira *et al.*, 2000).** Tomato plants are infested at any developmental stage, with females ovipositing preferentially on leaves. No tomato cultivars are entirely resistant to *T. absoluta*, but not all cultivars are equally susceptible **(Bogorni *et al.*, 2003; Oliveira *et al.*, 2009)**.

Although *T. absoluta* prefers tomato, it can also feed, develop and reproduce on other cultivated Solanaceae such as eggplant (*Solanum melongena* L.), potato (*S. tuberosum* L.) sweet pepper (*S. muricatum* L.) and tobacco, (*Nicotiana tabacum* L.) **(Vargas 1970; Campos 1976),** as well as on non-cultivated Solanaceae (*S. nigrum L., S. eleagnifolium L., S. bonariense L., S. sisymbriifolium Lam., S. saponaceum, Lycopersicum puberulum Ph.* etc.) and other naturally available host-plants such as *Datura ferox* L., *D. stramonium* L. and *N. glauca Graham* **(Garcia & Espul 1982; Larraı´n 1986a)**. On potato, *T. absoluta* only attacks aerial parts, thereby not directly impeding tuber development. Nevertheless, leaf feeding may indirectly lower potato yield and, under appropriate climatic conditions, *T. absoluta* could become a pest for the potato crop **(Eppo, 2005; Pereyra & Sa´nchez 2006).** Since the time of its arrival in Europe, additional plant species have been reported as alternative hosts. It has been reported in a Sicilian greenhouse of Cape gooseberry *Physalis peruviana* **(Tropea Garzia, 2009)** and has been found in Italy on bean, *Phaseolus vulgaris* **(EPPO 2009)** and on *Lycium* sp. and *Malva* sp. **(Caponero 2009).** Eggplant was reported to be the second-preferred host of *T. absoluta* after tomato; however, it is unclear if the species attacks only the leaves or if it attacks eggplant fruit. It has been reported on protected tomato and eggplant crops in a number of regions in Italy (**Ministero delle Politiche Agricole Alimentari e Forestali, 2009**). *T. absoluta* was also reported on greenhouse peppers and beans in Italy **(EPPO, 2009)**.

In South America, the preferred host of *T. absoluta* is tomato; the pest lays eggs in all aboveground portions of the plant (leaves, shoots and flowers) including on the fruit **(Vargas, 1970). Galarza, (1984)** showed that the tomato leaf miner is able to complete its development (from egg to adult stage) on *Solanum tuberosum*, *S. melongena*, *S. gracilius*, *S. bonariense* and *S. sisymbriifolium*, but development was interrupted (at larval instars I and II) on *Nicotiana tabacum* and *Solanum pseudo-capsicum*. **Cardozo *et al.*, (1994)**. Reported that *T. absoluta* is able to complete development on *Nicotiana tabacum*, and can use *Solanum elaeagnifolium* as an alternate host plant. **Fernandez & Montagne (1990)** they found that the tomato cultivar "Rome Gigante" was the preferred oviposition host and the best host for larval development, when compared to tomato variety Cerasiforme, eggplant, tobacco, *Solanum hirtum*, *Physalis angulata*, *S. americanum,* and potato.

In Europe and other parts of its expanded geographical range, *T. absoluta* prefers tomato. It can attack other solanaceous crops such as eggplant **(MPAAF, 2009; Viggiani *et al.*, 2009),** potato **(FREDON-Corse, 2009b ; Maiche, 2009),** and pepper **(MPAAF, 2009),** sweet cucumber (pepino) **(FERA, 2009b)** and Cape gooseberry **(Garzia, 2009b).** It was reported infesting common bean in Italy **(EPPO, 2009i; MPAAF, 2009).** Also, **Viggiani *et al.*, (2009)** reported that the tomato leaf miner causes direct and indirect damage to the production of tomato, potato, eggplant, and pepper, although it does not specify if the damage includes damage to fruit. The highest preferred hosts are listed as follow with minor information:

Black Night-Shade

Black night-shade belonging to family solanaceae. It is a common herb or short-lived perennial shrub, found in many wooded areas, as well as disturbed habitats. Characterized by their lack of prickles and stellate hairs, their white flowers and their green or black fruits arranged in an umbelliform fashion. It has been widely used as a food since early times and the fruit was recorded as a famine food in China.

Egg plant

Eggplant belonging to family solanaceae. It is known in South Asia and South Africa as brinjal. The eggplant is a delicate, tropical perennial often cultivated as a tender or half-hardy annual in temperate climates. Raw eggplant is composed of 92% water, 6% carbohydrates, 1% protein, and negligible fat 0.2%. It provides low amounts of essential nutrients, minor changes in nutrient compositions occur with season, environment of cultivation. Egypt is the 4th producers around the world 1.3 millions of tonnes /year.

Pepper

Pepper belonging to family solanaceae. The ideal growing conditions include warm soil, ideally 21 to 29 °C. Cultivars of the plant produce fruits in different colors, including red, yellow, orange, green, brown, white and purple. Peppers are rich sources of antioxidants and vitamin C. The level of carotene, like lycopene, is nine times higher in red peppers. Red peppers have twice the vitamin C content of green peppers. Egypt is the 8th producers around the world 475,000 tonnes /2007.

Tomato

Tomato belonging to family solanaceae. The tomato is native to western South America and Central America. Numerous varieties of tomato are widely grown in temperate climates across the world, with greenhouses allowing its production throughout the year and in cooler areas. In 2014, world production of tomatoes was 170.8 million tonnes, with China accounting for 31% of the total, followed by India, United States, Turkey and Egypt as the major producers. A tomato is 95% water, contains 4% carbohydrates and less than 1% each of fat and protein (table). In a 100 gram amount, raw tomatoes supply 18 calories and are a moderate source of vitamin C (17% of the Daily Value), but otherwise are absent of significant nutrient content. A tomato is the main host plant for many different insect species.

Potato

Potato belonging to family solanaceae. Raw potato is 79% water, 17% carbohydrates (88% of which is starch), 2% protein, contains negligible fat (table). In a 100 gram amount, raw potato provides 77 calories and is a rich source of vitamin B6 and vitamin C (23% and 24% of the Daily Value, respectively), with no other nutrients in significant amount. When a potato is baked, contents of vitamin B6 and vitamin C decline with little significant change in other nutrients. The world dedicated 18.6 million hectares in 2010 for potato cultivation producing about 17.4 tonnes/hectare. The optimum cultivation conditions spring weather with temperatures ranged from 15 to 20 °C.

Tobacco

The plant is part of the genus *Nicotiana* and of the solanaceae family. Tobacco is cultivated similarly to other agricultural products. It is cultivated annually in cold frames or hotbeds, as their germination is activated by light and can be harvested in several ways. Tobacco production requires the use of large amounts of pesticides. Several species of insects pose serious threats to tobacco in the field, the greenhouse, and the curing barn. Insects damage the roots, destroy the leaves and buds, reduce leaf quality, and transmit several important tobacco diseases as tobacco budworm, Hornworm, tobacco flea beetle, green peach aphid, wireworms, cabbage looper, green stink bug and leaf miners. China is the 1st producers in the world.

Datura

Datura is a genus of nine species of poisonous vespertine flowering plants belonging to the family Solanaceae. It is herbaceous, leafy annuals and short-lived perennials which can reach up to 2 m in height. Datura species are usually planted annually from the seed produced in the spiny pods, but with care, plants can be overwintered. Most species are suited to being planted outside or in containers. As a rule, they need warm, sunny places and soil that will keep their roots dry.The larvae of some Lepidoptera (butterfly and moth) species, including *Hypercompe indecisa*, eat some Datura species.

Cape gooseberry

Cape gooseberry is a plant species of the genus *Physalis* and belonging to family Solanaceae. Widely introduced in the 20th century, Cape gooseberry is cultivated or grows wild across the world in temperate and tropical regions but it is an annual in temperate locations and perennial in the tropics. In South Africa, cutworms attack the cape gooseberry in seedbeds, red spiders in the field, and potato tuber moths near potato fields. Hares damage young plants, and birds eat the fruits. Mites, flea beetles, whitefly, and flea beetles can be a problem. Powdery mildew, soft brown scale, root rot and viruses may affect plants. In New Zealand, plants can be infected by *Candidatus liberibacter* .

Bean

A bean is a seed of one of several genera of the flowering plant family Fabaceae, which are used for human or animal food. Beans are a summer crop that need warm temperature to grow. Maturity is typically 55–60 days from planting to harvest. Beans are high in protein, complex carbohydrates, folate, and iron. Myanmar country is the 1st producer in the world. Although numerous insect pests attack all parts of beans, bean stem maggots and bruchids are the most important field and storage pests, respectively. Foliage beetles, flower thrips, pollen beetles, pod borers, pod bugs, and sap suckers such as aphids also inflict significant damage.

Malava

Malva is a genus of about 25–30 species of herbaceous annual, biennial, and perennial plants in the family Malvaceae. The genus is widespread throughout the temperate, subtropical and tropical regions of Africa, Asia and Europe.

Lycium

Lycium is a genus of flowering plants in the family of Solanaceae. The genus has a disjunct distribution around the globe, with species occurring on most continents in temperate and subtropical regions.

Lycium are shrubs, often thorny, growing 1 to 4 meters tall. The leaves are small, narrow, and fleshy, and are alternately arranged, sometimes in fascicles. Many insect pests attacks this plant as western flower thrips, green peach aphid, brown-winged green bug, plant hopper, potato ladybeetle, gelechiid moths and cotton bollworm.

The tomato leaf miner *Tuta absoluta* in Egypt

Agriculture in Egypt

Egypt is currently facing many challenges for the provision of human requirements of food along with the sustainable development. Agriculture is the main sector responsible for providing food and raw materials for most of industries responsible for the sustainable development in Egypt.

In the second of the 20th century many people in different parts around the world suffer from acute shortage of food and from malnutrition. Insect pests, plant diseases and adverse climatic conditions are the main reasons in this shortage of productivity of crops. So many programs focused on the utilization of the protected agricultures and glass houses for increase of food requirements and for resolving the problems of low productivity of some crops especially vegetable crops.

Egyptian agriculture is characterized by the production of many exported vegetable crops such tomato, potato, cucumber and green beans ....etc, and the expansion in cultivation of these crops is the best agricultural investment in many countries of the world due to the shortage or reduction of the life cycle and recultivation of them more than once throughout the year compared with permanent crops, also the daily requirements of vegetable crops by consumer increased the economic importance of these crops at the investor.

Agriculture in Egypt is the main and only sector responsible for the food viability and it is one of the most important development axes of the Egyptian economy. The agricultural area in Egypt's was 15.8 million feddans in 2010, and the expansion of agricultural investment led to increase of this area. Vegetables are regarded as strategic crops, which provide population with food security, being an essential component of daily food requirements. Consumption of vegetable crops has been escalating over the last decades due to growth of population and upgraded health and nutrition awareness of citizens. the expansion in cultivation of these crops is the best agricultural investments in many countries of the world due to the shortage or reduction of the life cycle and regeneration of it is cultivations more than once throughout the year compared with permanent crops, also the daily requirements of vegetable crops by consumer increased the economic importance of these crops at the investor, so the interest in the expansion of cultivation of these crops was increased.

Egyptian agriculture is characterized by the production of many exported vegetable crops such as tomato, potato, cucumber and green beans ....etc. The annual production of crop in Egypt is 9204097 tons of tomato fruits from about 9000 ha of the cultivated area (**Moussa *et al.*, 2013**). So, it is considered as the 5th largest tomato producers in the world (**WPTC, 2011**). Also, it is considered as 7th in the world in cultivation and production of beans. Different vegetable crops as potato, pepper and cucumber are the most important vegetable crops in Egypt cultivation and production and export (**Publisher: Economic Affairs sector, Egyptian Ministry of agriculture**).

Because of the importance of economic and export of these crops, so it becomes necessary for us improvement of the quality and quantity of their production which qualifies for placement on lists of export crops and through the development of organic agriculture programs for the production of safe food without chemicals, whether pesticides or fertilizers.

Due to the shortage of the life cycle of vegetable crops and regeneration of its cultivations, more than once through the year, it represent the main host plants to many different insect pests affecting crop quality and productivity such as leaf miner insects (*Tuta absoluta* & *Liriomyza tifolii*) the larvae of these pests attack different vegetable crops as tomato, potato, eggplant, pepper and bean in all developmental stages, where the larvae prefer leaves, stems, flowers and especially fruits besides mining their leaves which can affect the plant photosynthetic capabilities causing high reduction in productivity 80-100% (**FERA, 2009**).

All the aforementioned insect pests are the most important reasons for low productivity of most vegetable crops especially tomato plants. So, this article focus as a reference background on *T. absoluta* in Egypt.

Distribution of *Tuta absoluta* in Egypt

Egypt, lies in the subtropical region, is considered one of the important tomato producers in the world [according to Bulletin of World Processing Tomato Council (WPTC) in 2011] that has an appropriate climate for tomato throughout the year in three different plantation seasons viz., winter, autumn and summer. Annually, it produces about 9,204,097 tons of tomato fruits from about 9,000 ha of cultivated area. Tomato crop is one of the most important vegetable crops in Egypt and is considered as the fifth largest tomato producer in the world but one of the world's countries that suffer from *T. absoluta* population increase **(Reda and Hassan, 2015).**

T. absoluta introduced and moved from border of Libia to inside Egypt in 2009 and distributed in all Egyptian governorates up to 2011. Overlapping periods of planting tomato within the same year in addition the climatic conditions, allow the favorite tomato host plant available all year round. The last allowed quick movement of this insect pest. Egypt is planting tomato 4-5 times /year.

Tomato crop is the first vegetable crop in Egypt; it is grown in four rotations and covers about 3% of Egypt's total planted area of newly reclaimed and old lands under greenhouses and in outdoor areas. FAO classified Egypt during 2006-2008 as the 5th country around the world and second around the Mediterranean countries in production and exportation of tomato.

According to the Egyptian Ministry of Agriculture, the tomato cultivations distributed in all Egyptian governorates with the largest areas located in Beheira, Fayoum, Sharkia and Qena governorates. **Tamerak S., (2011)** recorded *T. absoluta* in Egypt at the border of Libya in the second half of 2009 and stated that it had reached to Giza and was established in all governorates of Egypt and reached to the border and northern part of Sudan on June 2011.

Table : distribution of *T. absoluta* in Egyptian governorates

Governorate	% establish		
	2009	2010	2011
Marsa-Matrouh	40	80	100
Alexandria	-	70	100
Behera	-	60	100
Sharkia	-	50	100
Qulupia	-	50	100
Dimitta	-	40	100
Dakahlyia	-	40	100
Ismailia	-	30	100
Kafer- El Sheikh	-	30	100
Giza	-	20	100
Fayoum	-	-	100
Bani – Sweif	-	-	100
Minia	-	-	100
Assiut	-	-	100
Sohage	-	-	100
Qena	-	-	100
Aswan	-	-	100

Many trials estimated the annual generations of *T. absoluta* under Egyptian climatic conditions. . Population of the *T. absoluta* in Qena governorate gave 13 generations while in El-Beheira, Giza and Fayoum governorates gave 11,12 and 12 generations respectively under current climate conditions. In future climatic conditions (in 2050 and 2100) number of insect generation will increase to 12 and 13 generation, respectively in El-Behera governorate. Numbers of generation will increased with same rates in rest governorates **Abolmaaty et al., (2010).** In Baltiem district, Kafrel-Sheikh Governorate, Egypt, male moths was increased gradually to reach the reliable occurrence of the 1st generation during the last week of March in spring plantation and first week of May in summer plantation. The peak of this generation was recorded on March, 26th and May, 3rd. After this period the reliable occurrence of the 2nd generation, the pest took place in the first week of April and first May to reach its peak on April, 4th and May, 8th **(khider *et al.*, 2013).** Using light traps, **Salama *et al.*, (2014)** documented eight peaks of *T. absoluta* during the year 2012, including three peaks in early summer plantations and four peaks in nile plantations and an additional peak between the two rotations. In 2013, five peaks were recorded in the summer plantations and three peaks in winter plantations. The rate of increase in insect population reached a high index in March-April when the average temperature ranged between 16.19 and 21.13 °C. Also, **Mahmoud *et al.*, (2015)** reported that in both winter and summer plantation the population of *T. absoluta* started to appear with the first day of tomato plantation and increased gradually with increasing the development of tomato plant till flowering stage at which the peak of population was noticed using white pheromone traps baited with pheromone lure type TUA-500 and indicated that, in winter plantations, *T. absoluta* population started to appear in the 2nd week of Jan., and increased gradually forward till reached its maximum activity in the 3rd week of Feb., the 3rd week of March and the 1st week of April then decreased and approximately declined in the 4th week of April. *T. absoluta* had four peaks during winter season; the first peak occurred in the 4th week of Jan., the second peak occurred in the 3rd week of February, the third was in 3rd week of March while the fourth peak occurred in the 1st week of April. In summer plantations, the population of *T. absoluta* started very weak, in the 4th week of June. An approximately outbreak occurred in the 1st week of Sept. and the population then decline in the 2nd week of Oct. Also, four peaks of *T. absoluta* were recorded in summer plantation, the first peak was in 2nd week of July, the second was in 1st week of August, the third and fourth were in the 1st and the 4th week of Sept., respectively. Results **Reda and Hassan, (2015)** reflected that male population density of *T. absoluta* varied from season to another. Spring season was the highest followed by summer, but each of winter and autumn were the lowest, that there were no significant differences between the last two seasons. Effect of tested climatic factors is obvious along a year, not can be observed in specific seasons, that their combined effects responsible as a group for 34.09% and 35.76 % on population density in the both years of study, respectively. Eleven annual generations were observed along a year in both years of study, moreover times and duration of all estimated generations were paralleled in the both used mains of generation estimations. The first and eleventh generations were nearly longest but with lowest in insect male population numbers. The Ninth generation, considered as overlapped generation between summer and autumn season. Each of spring, summer and autumn season have three generations. **Al-Sawy *et al.*, (2015)** Observed the population fluctuations of *T. absoluta* during two subsequent tomato seasons 2014 and, 2015. Direct examination of tomato leaves revealed a consistent fluctuation pattern of both the insect larvae and the true spiders in the two seasons. Insect larvae peaked on the mid of June having (65 larvae/10 plants) while the true spiders peaked one week later having (5 spiders/10 plants) in both seasons, respectively. The observed insect larvae population reduction accompanied with an increase in true spiders population during the midseason may reveal a possible predation relationship between them. The afterward detected declining of both populations may be attributed to the initiations of plant dryness prior to the end of the season. The current study strongly emphasizes the importance of pheromone traps as an effective and powerful predicting method for early pests warning.

In the same line, the population fluctuation of *T. absoluta* using mass trapping in Qaha (Qalyoubia governorate, Egypt) during spring-early summer cycle in 2013 and 2014 was recorded by **Abd El-Ghany *et al.*, (2016)** depending on the catching number of moths/week during the plantation period. Results of the first season indicated gradually increasing from the 4th week of April until an approximately outbreak on 1st week of May. Subsequently, two peaks recorded in 3rd week of May and 1st week of June. The population decline gradually from the 2nd week of June and approximately reaches the lowest in the 3rd week of July. However in 2014, *T. absoluta* had five peaks during the plantation period. Two peaks were recorded in April; the first peak occurred in the 2nd week, and the second peak occurred in the 4th week. Subsequently, the third peak was recorded in the 3rd week of May. Additionally, two peaks were recorded in the 1st and 3rd week of June. Moreover, the effect of trap color (red and yellow) was evaluated by comparing weekly catching number of moths/trap. A glance in data that indicate that, red traps are more attractive than the yellow traps.

T. absoluta is one of the most important lepidopterous insect pests on tomato. It prefers to lay eggs on tomato leaves and after egg deposition; larvae penetrate the leaves, stems or fruits causing mines and galleries. No tomato cultivars are entirely resistant to this insect, but not all cultivars are equally susceptible. Similar observations were recorded by **Borgorni *et al.*, (2003). Shehata *et al.*, (2016)** evaluate the susceptibility of different strains of tomato during the early summer and nile rotations in 2012 and also during summer and winter rotations in 2013. It appears that the native strain was more susceptible to infestation compared to the strain GS and super hybrid in the early summer rotation of 2012. In the nile rotation, the strain 77 was more susceptible to infestation as compared to the strain 010 where the number of eggs, larvae and mines on the leaves of the strain 77 were significantly higher compared to that of the strain 010. In the summer rotation of 2013, the strain 2243 appears to be more susceptible to infestation with *T. absoluta* as compared to the strain GS. On the other hand, the strain coded 5656 showed to be more susceptible to infestation as compared to the strain 010, judged from the number of eggs, larvae and mines (tunnels) recorded during this season on the leaves of this strain. These findings clearly indicate the variation in the susceptibility of the most dominant tomato strains cultivated in Egypt. While, **Ata and Megahed, (2014)** studied the susceptibility of other tomato strains in Egypt and showed that the tomato plant variety of Alisa is more susceptible to the infestation with *T. absoluta* larvae than the other variety (H.S.S.), the general mean number of larvae /leaf was 3.3 and 2.8 /leaf in Alisa and H.S.S varieties; respectively. While, the general mean numbers of mines/leaf were 5.5 and 4.0 in Alisa and H.S.S varieties; respectively. The infestation percentage of leaves take the same trend with mean numbers of mines and larvae/leaf, it was very low in March as 0.9% and 0.2% in Alisa and H.S.S varieties respectively, and it was gradually increased reaching to 100% and 97.6% of infestation in Alisa and H.S.S varieties, respectively at the end of the season.

Seasonal abundance of this pest was studied in Qaha region, Qalyubiya Governorate, on four tomato cultivars {Alissa F1, Super strain B, G.S 12 F1 and Logain (E603 F1)} during early and late summer plantations throughout years 2013 and 2014 by **El- Badawy *et al*., (2017)** and recorded that the population density was higher in early summer and summer plantations of year 2013 than that recorded in year 2014. The highest seasonal mean number was recorded on Alissa cultivar for both studied years followed by Super strain cultivar then G.S. cultivar. While the lowest mean number was achieved with Logain cultivar. Susceptibility interpretation of Logain and Alissa tomato cvs. to *T. absoluta* may be attributed to the presence of high contents of the toxic and repellent hydrocarbons octacosane and hexacosane in Logain tomato cv. and high content of the attractant hydrocarbon tetracosane in Alissa cv so it can be recommend to use the tolerable Logain tomato cultivar in breeding programs and also, preparing a commercial product/formulation from hydrocarbons octacosane and hexacosane to be used as repellent and tetracosane as a trap to *T. absoluta*. The difference in the *T. absoluta* preference between the different tomato cultivars can be attributed to differences in leaf volatile compounds (**Proffit *et al*., 2011; Cherif *et al*., 2013).** Indeed, based on oviposition bioassays,

Proffit ***et al*****,. (2011)** demonstrated that *T. absoluta* females laid more eggs in response to cvs. Santa Clara and Carmen as compared to cv. Aromata. The same authors found that overall leaf volatile composition of cv. Aromata differed significantly from cvs. Santa Clara and Carmen, due to differences in proportions of minor compounds and due to the absence of several compounds, mostly terpenes, in cv. Aromata. **De Oliveira** ***et al*****., (2012)** showed that oviposition rate and damage on plants were significantly lower on tomato strains rich in one of the following allelochemicals: 2-tridecanone or zingiberene.

Determination of the seasonal variation and population density of *T. absoluta* in tomato cultivations in Bernucht, Giza governorate for two years, starting from the last week of January 2012 till January 2014 was recorded by **Salama** ***et al*****., (2014)** using light traps for catching the adult moths and reported eight peaks of *T. absoluta* as illustrated in the following figures (6 & 7).

Data illustrated in **Fig. (6)** appears that *T. absoluta* has 8 peaks around 2012. Three peaks in early summer rotation, the first peak occurred in the 3rd week of March, the second occurred in early May, the third in the first week of June. Also, one peak was recorded in mid of July (between early summer and nile rotations) and four peaks of *T. absoluta* were recorded in the nile rotation. The first peak was in early September, the second was in mid of October and the third was in early December, while the fourth peak began in January 2013.

While, the data illustrated in **Fig. (7)** noticed that, *T. absoluta* has five peaks in summer plantation (from March to the first of August), the first peak occurred in the first week of March, the second in the mid of April, the third in the mid of May, the fourth in early July and the fifth peak was recorded in the first of August. Also, three peaks of *T. absoluta* were recorded in winter plantation. The first peak was in early September, the second was in early of December, the third was in the mid of January of 2014.

Fig. 6. Population fluctuation of *T. absoluta* using the light trap in Giza governorate during 2012-2013.

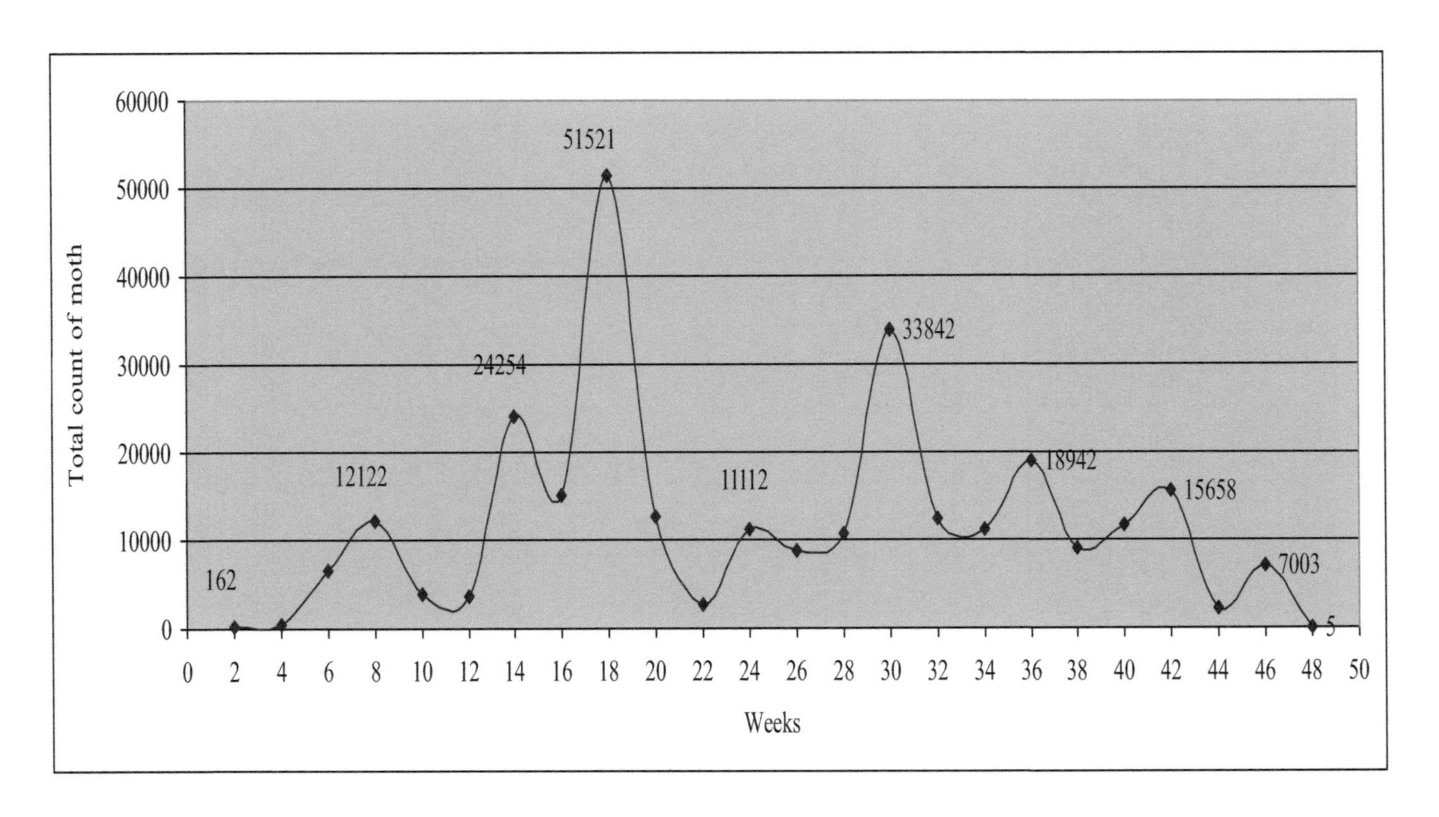

Fig. (7): Population fluctuation of *T. absoluta* using light traps in Giza governorate during 2013-2014.

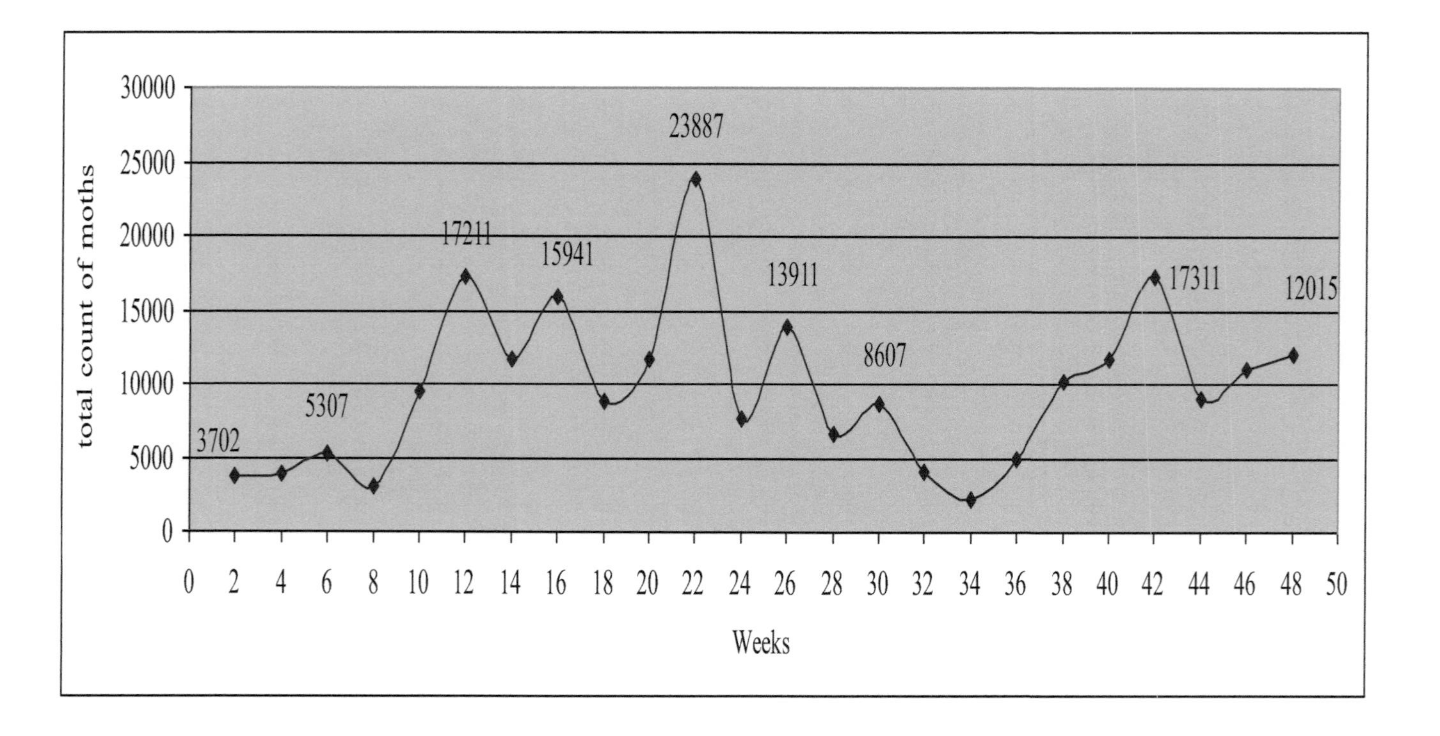

Field observations of **Shehata (2015)** clearly indicated that all different stages of *T. absoluta* are present all the year in the samples collected from tomato plants and light traps collections. This suggested that the insect emergence occurs throughout the year. The total effective temperatures (thermal constant) represent the limiting and required factor for development of each stage. In order to determine the thermal threshold (below which development was inactivated) and thermal constant for each of the insect developmental stages, the formula of **Bodenheimer (1951)** and **Jasic (1975)** were applied. The thermal threshold was calculated first using the equation, $y = n (t - x)$,where **y** equals the thermal constant, **x** is the thermal threshold or the zero of development of the stage below which development of any stage was inactivated and **n** equals the average duration of development at the average temperature **(t)**. By applying the same formula, the thermal constant was calculated. So, the calculation of the insect thermal requirements helps in the prediction of the pest occurrence and plays a good role in the pest management programs.

Based on accumulation of the thermal developmental requirements **Shehata (2015)** expected the frequencies of *T. absoluta* moth emergence in the field in different Egyptian governorates (Giza, Qena and Mersa- Matrooh) depending on variation in climatic conditions in these governorates.

So, the investigations can help in predicting *T. absoluta* annual generations and expected times of moth frequency in the field under current and expected future climate changes by using the relationship between the accumulated thermal heat units expressed as degree days and its population fluctuation in the experimental area. It can be easily detected and help in designing program for pest management.

Taking in consideration that the accumulation of heat units started on 1st of February 2012, it appears that the cycles for the insect emergence occur throughout the year and the termination of pupal development at any time represents the expected time for the moth emergence. The data clearly indicate that the first cycle of adult emergence extends through 20.82 days in February. The second cycle begins in the remaining 8.18 days of February to give 39.3% of adult emergence. This cycle extends through the first 9.98 days in the following month (March) to attain the remaining 60.7 % of adult emergence and to complete the 2nd cycle. Again the 3rd cycle extends through within the same month. This phenomenon was continued until the end of January 2013, when the 36th emergence cycle was detected after 22.41 days of January 2013.

As already mentioned, variation in climatic conditions in different governorates certainly affect the insect development and it follows that insect generations will differ. The duration of the insect generation reared at 26.03 °C was 38.49 days compared to 79.5 days when reared at 15.32 °C. So, the calculated developmental zero was 5.268°C and the thermal constant was 799.1 DDs. So, the expected frequencies of annual generations was determined in three governorates Giza, Qena (one of the largest cultivated tomato areas) and Mersa-Matrooh (along the Mediterranean cost).

In Giza governorate, the life cycle duration of the insect was 91.52 days in February 2012 when the average temperature was 14°C, this decreased gradually in the following months and the shortest duration was 33.92, 32.31 and 32.37 days in June- August 2012. This was followed by another increase and the longest duration was recorded in winter months. The average life cycle duration from February 2012 to January 2013 was 52.44 days when the mean annual temperature was 22.8 °C and the number of annual generations was **8.05**. Analysis of the data indicated that expected time of the first generation was on 11.66 days in April 2012. This cycle passed through the 18.34 days of April where the rate of generation development was 0.4. This cycle extended through the first 22.63 days in May to attain 0.59 of generation and to complete this generation cycle on 22.63 May 2012. This phenomenon was continued until the end of January 2013, when the 8th generation cycle was obtained after 28.2 days of January 2013. Repeating these investigations during 2013 in Giza governorate (the mean annual temperature, 22.2 °C), reveal that the number of annual generations was **7.8**. Applying the same calculations in Qena governorate (the mean annual temperature, 26.04 °C) indicate that the number of annual generations during 2012 was **9.52**. This is certainly correlated with the high temperature records in this governorate. In Mersa-Matrooh governorate, with low temperature records (the mean annual temperature, 20.53°C), the number of the annual generations was found to be **7** during 2012.

This approach has been adopted by previous authors with various insect species e.g. **Sevacherian *et al.,* (1977); Farag *et al.,* (2009)** stated that accumulated thermal units have been used to predict the seasonal development and emergence of various insects. **Vercher *et al.*, (2010)** reported 10-12 generations per year in South America. While, **Varges, (1970)** found 7-8 annual generations in Chile. **Barrientos *et al.*, (1998)** stated that the thermal constants were 103.8 ± 1.4, 238.5 ± 0.5 and 117.3 ± 5.3 DDs for eggs, larvae and pupae of *T. absoluta*, respectively, while the total thermal constant from egg to adult was estimated to be 453.6 ± 3.9 DDs. **Abolmaaty *et al.*, (2010)** recorded that the population of *T. absoluta* in Qena gave the highest number of generations as compared to other governorates. Mean thermal units for complete development in Qena was 471 units. **Hamdy, (1990a)** stated that the expected date for generations of the oleander scale, *Aspidiootus hederae*.

The moths after emergence will be able to detect the source of nectar, either from the field where they emerge or from flowering weeds on the borders of cultivations or from other host plants within the limit of their activity in the neighboring. Also, evaluation of the flight range is very important to predict with the infestation in the other healthy adjacent fields. The ability of the *T. absoluta* moth to fly for a distance of 0.4 kilometers overnight is of great importance in regulating its dispersion and oviposition on different host plants during its life span as recorded by **Salama *et al.*, (2015).**

Table of contents

FSC
www.fsc.org
MIX
Papier aus ver-
antwortungsvollen
Quellen
Paper from
responsible sources
FSC® C141904

Druck:
Canon Deutschland Business Services GmbH
im Auftrag der KNV-Gruppe
Ferdinand-Jühlke-Str. 7
99095 Erfurt